SMP 11-16

Book A3

CAMBRIDGE
UNIVERSITY PRESS

Published by the Press Syndicate of the University of Cambridge
The Pitt Building, Trumpington Street, Cambridge CB2 1RP
40 West 20th Street, New York, NY 10011-4211, USA
10 Stamford Road, Oakleigh, Melbourne 3166, Australia

First published 1995

Printed in Great Britain at the University Press, Cambridge

A catalogue record for this book is available from the British Library.

ISBN 0 521 47842 1 paperback

Acknowledgments
Photographs: Cover, 4, 5, 6, 7, 11, 12, 13t, 13b, 14t, 15, 16, 18, 19, 23,
29t, 29m, 29bl, 30 (bottom four) 31b, 34, 35, 38, 41, 42, 45, 46
Graham Portlock; 32, 33 Sally and Richard Greenhill; 30t Angelo
Hornak Library; 31 B. T. Batsford Ltd; 29br, 9 (Sneezeweed)
Wildlife Matters; 13m Huss Maschinenfabrik; 8, 9 A–Z Botanical
Collection; 14m Paul Scruton.

Illustrators: 8, 17, 20t, 21t, 24, 25, 26, 27, 40t, 40m Debbie Hinks; 10
Dale Edna Evans; 11, 28, 35tr, 36tl, 36bl, 37l, 37r Sue Hellard; 14, 16,
20m, 35tl, 37m, 40b, 48m Martin Sanders; 20b, 21b, 39r, 41, 42m, 43l,
43b, 45m, 47t, 47m Peter Kent; 22, 48t Martin Ursell; 34, 35b, 36tr,
36br, 39l, 42 (scales), 43 (scales), 44, 45b, 47b, 48b, 49 Jeff Edwards.

Handwriting: Hilary Evans.

The author and publisher would like to thank the pupils and staff of
Impington Village College for their help and co-operation.

Contents

Rulers	**4**
Gardens	**8**
Halving recipes	**11**
Horizontal	**12**
Angle art 1	**15**
Money	**16**
Playing Scrabble	**18**
a.m. and p.m.	**20**
Calendar	**22**
Angle art 2	**23**
Cafe	**24**
Make a kilogram	**28**
Vertical	**29**
The Hancock Family	**32**
Honeycombs	**34**
We asked 100 teenagers	**35**
Four cubes	**38**
24-hour times	**39**
Kitchen scales	**41**
Review: Book A2 and pages 4 to 17	**44**
Review: pages 11 to 22	**45**
Review: pages 15 to 27	**46**
Review: pages 24 to 37	**47**
Review: pages 32 to 43	**48**
A time to remember	**49**

For this book to be effective, it needs to be supported by

- teacher-led classroom activities that highlight the mathematical ideas
- teacher-led discussion of what pupils see and read on the page
- checking by the teacher that pupils have understood key ideas together with structured help and practice when they have not.

The symbol shows where such teacher input is specially important. Advice on how to provide it is marked with the same symbol in the Teacher's guide.

Rulers

Things to talk about

- Go round your classroom and find at least three rulers.
  Look for rulers that are different from one another.

- Which of your rulers is the longest?

- Which ruler is the shortest?

- Look at the numbers on your rulers.
  Which ruler has most numbers on it?
  Which ruler has the **biggest** number on it?

A Measuring in centimetres

The numbers on this ruler are for
measuring lengths in centimetres (cm).

A1 How long is the pencil?
Write your answer with **cm** after it.

A2 Pick out a ruler that has a scale like this one.
Use it to find how long these lines are.

(a) ————————————————————————

(b) ————————————————

(c) ——————————————————————————

(d) ————————————————————

(e) ——————————————————————

B Centimetres and millimetres

These small marks divide each centimetre into 10 parts.
Each of the parts is called a millimetre (mm).
Check that there are 10 millimetres in
one centimetre on the ruler.

This nail is 5 centimetres and 3 millimetres long.
We can write its length as **5 cm 3 mm**

Get worksheets A3–1 and A3–2 .

Measure the scissors on worksheet A3–1 in centimetres and millimetres.
The arrows show you where to measure.

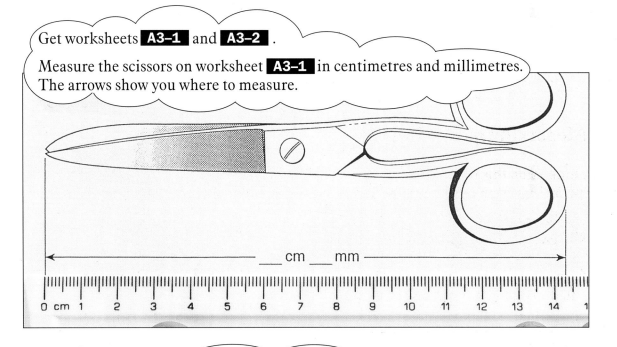

_____ cm _____ mm

Write the length in the blank spaces on the worksheet.

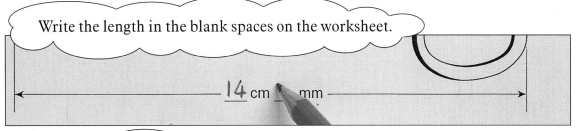

14 cm _____ mm

Measure the other lengths the same way.

5

C Just using millimetres

- Find a ruler with a scale like this one.
  What length are these small parts?

- Why do the numbers go up in tens like this?

You can measure the whole length of something in millimetres.
The nail is 53 millimetres long.
We can write this as **53 mm**.

Worksheets **A3–3** and **A3–4** have the same objects as before.
But now you have to measure them in millimetres.
Fill in your answers on the worksheets.

D Inches

These rulers have scales marked in **inches**.

An inch

An inch is roughly
the length of the top part
of your thumb.

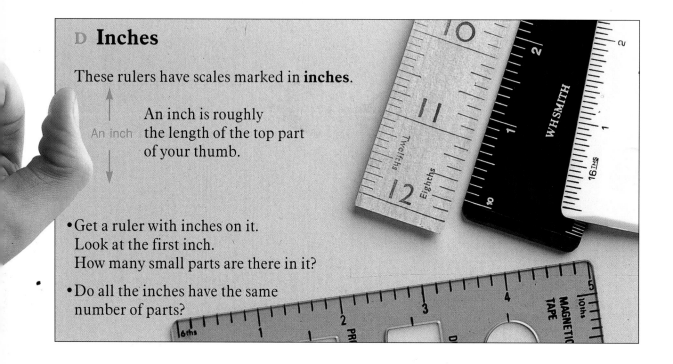

- Get a ruler with inches on it.
  Look at the first inch.
  How many small parts are there in it?

- Do all the inches have the same
  number of parts?

E Finding the 0 mark

When you measure something, one end of it goes
against the 0 mark on the ruler.

The 0 mark on an ordinary plastic ruler is a short distance in from the end.
Sometimes the 0 is not written on.

These rulers and measuring tapes are from a workshop.
Show your teacher where the 0 mark should be on each one.
Explain how you decided.

Gardens

A Choosing flowers

You need: felt tip pens or crayons
scissors

Get worksheet **A3–5**.
Colour the flower pictures.
Use these photographs to get the colours right.
Cut out the flower labels.

Get worksheet **A3–6**.
The dotted lines divide the flower beds into sections.
Decide which flowers you will put in each section.

Think about: the heights of the flowers
whether they need sun or shade
which colours go together
when the flowers bloom

Move the flower labels around to help you decide.

When you have decided, write the names of
the flowers where you want them to grow.

Colour the sections to show how
the garden will look **in May**.
If a flower does not bloom in May,
colour its section green.

Get another copy of worksheet **A3–6**.
Keep the flowers where they are, but show
the colours **in August**.

Did the garden look the way you wanted it?

Bishop's hat

Bleeding heart

Cyclamen

Globe thistle

Jacob's ladder

Ligularia

Sneezeweed

Bluebell

Catmint

Crocosmia

Daffodil

Evening primrose

Forget-me-not

French honeysuckle

Japanese anemone

Lady's mantle

Leopard's bane

Speedwell

Wallflower

White everlasting

B Design a garden

Get some plain paper.
Design a garden of your own with flower beds.
Your garden can have some of these features if you like.

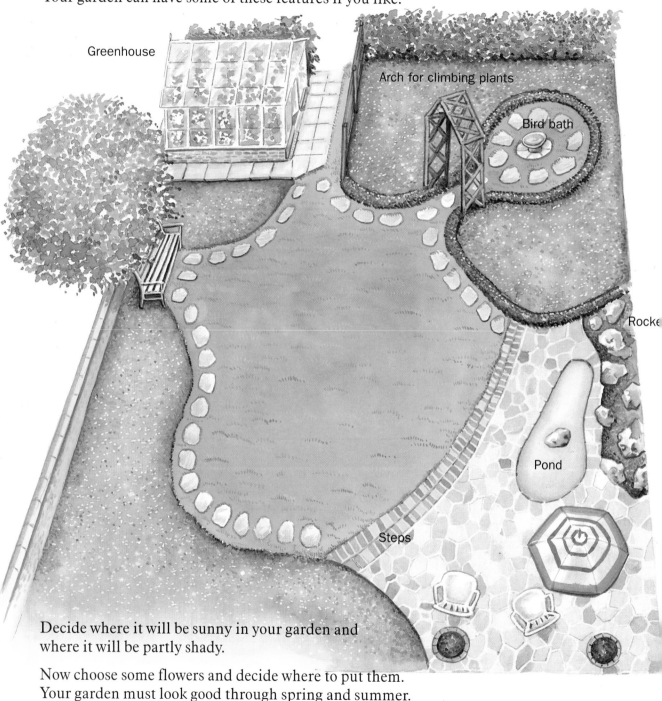

Greenhouse

Arch for climbing plants

Bird bath

Rocke

Pond

Steps

Decide where it will be sunny in your garden and
where it will be partly shady.

Now choose some flowers and decide where to put them.
Your garden must look good through spring and summer.
You can use the flowers on pages 8 and 9,
or you can find out about other flowers and use them too.

Halving recipes

Apple crumble

Enough for 6 people

For the **apple mix** you need: 2 large cooking apples
50 g sultanas
50 g sugar
a pinch of cinnamon

Peel the apples and chop them up.
Put them in a baking dish and sprinkle the sultanas,
sugar and cinnamon over them.

For the **crumble mix** you need: 200 g flour
100 g butter
80 g sugar

Mix everything together until it is crumbly.

Spread the crumble mix over the apple mix and
bake it in an oven for 40 minutes at 375°F (190°C).

Tony wants to cook apple crumble.
But he only wants enough for 3 people, not 6.
So he decides to **halve** the amounts.

1 Write the amounts he will need for the apple mix.

2 Write the amounts he will need for the crumble mix.

Get the recipe for something that **you** like to make or to eat.

3 How many people is the recipe for?

4 If you halve the recipe, how many people will there be enough for?

5 Write the amounts for the halved recipe.

Horizontal

A Getting things level

You need: a ping-pong ball, marble or round coin
a spirit level

You can use a ping-pong ball to check
whether the top of a table is level.
Put it gently on the table.
If it stays still, the table is level.

If you don't have a ping-pong ball you can use a marble,
a round coin or something similar.

Your teacher will give you some activities to do.
They are all about checking whether a table is level.

B Using a spirit level

A surface that is flat and level is called **horizontal**.

If you put a spirit level on a surface,
the position of the bubble tells you
one of these things.

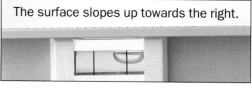

The surface slopes up towards the right.

The surface slopes up towards the left.

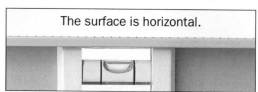

The surface is horizontal.

Rebecca uses a spirit level to test some surfaces in her classroom.
She makes three lists like this.

Slopes up towards the right	Slopes up towards the left	Horizontal
The top of the filing cabinet	The window ledge	My desk
The floor under the blackboard	The top of the door	

Activity

Test some surfaces in your classroom with a spirit level and
make three lists like Rebecca's.

These bricks are horizontal,
even though the wall is on a hill,
The builder probably used a spirit level.

This Arabian carpet
gives you a frightening ride,
but it stays horizontal.

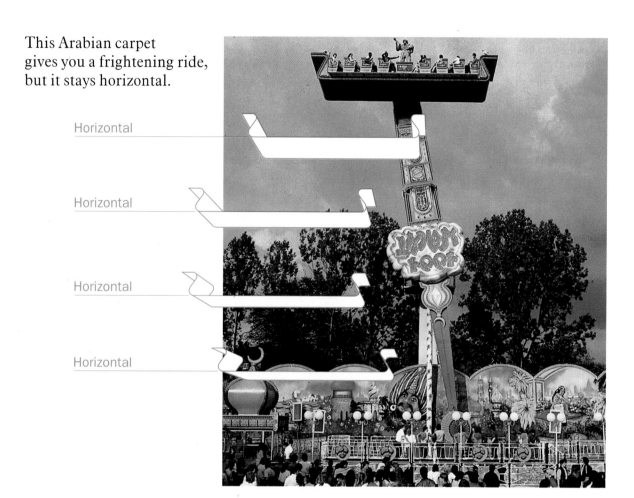

Horizontal

Horizontal

Horizontal

Horizontal

The surface of a pool table has to be **exactly** horizontal.
You adjust the height of each leg by turning a screw.

C Looking at liquids

C1 Put some water in a beaker.
What happens to the surface of the water as you tilt the beaker?

C2 Ratna has made this jelly.
How did she do it?

C3 Look at the photograph on the front cover of this book.
How was it taken?

C4 This 'flat' roof actually has a slight slope.
Why is this?
Think of some other surfaces that deliberately have a slight slope.

Angle art 1

patterns that make a good wall display

You need compasses, an angle measurer and felt-tip pens.

1 Put a dot near the centre of
a piece of plain paper.
Set your compasses to 4 cm.
Draw a circle with its centre on the dot.

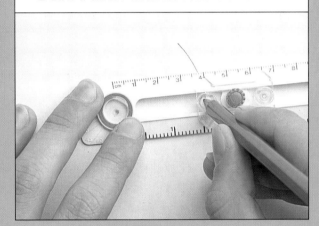

2 Put the centre of the angle measurer
on the dot.
Mark off every 10 degrees around
the angle measurer.

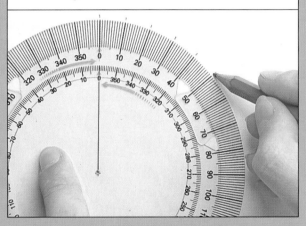

3 Draw a line from the centre to each of
the marks you have just made.
Continue each line to
the edge of the paper.

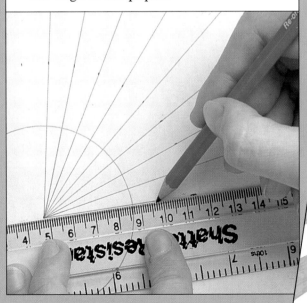

4 Choose your own colours to make
this kind of pattern.

15

Money

You need bank notes, scissors and crayons or felt-tip pens.

We use coins for small amounts of money.

We use notes for large amounts.

1 Which do you use more often, coins or notes?

2 Which are the more hard-wearing, coins or notes?

Notes have features that make them difficult to copy.

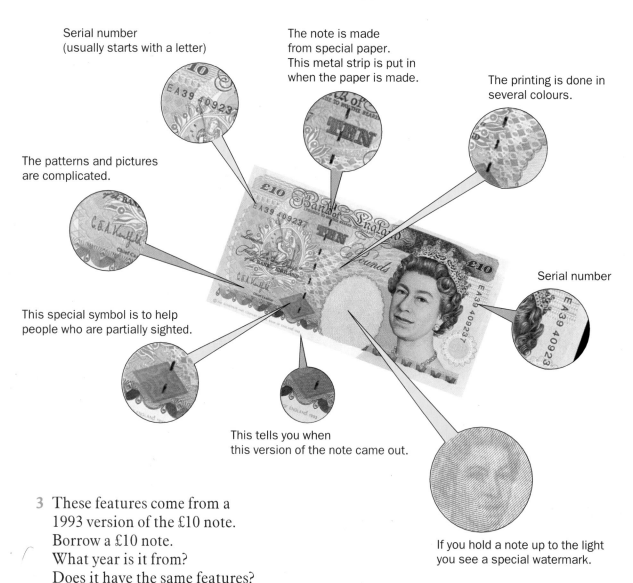

Serial number
(usually starts with a letter)

The note is made
from special paper.
This metal strip is put in
when the paper is made.

The printing is done in
several colours.

The patterns and pictures
are complicated.

Serial number

This special symbol is to help
people who are partially sighted.

This tells you when
this version of the note came out.

3 These features come from a
1993 version of the £10 note.
Borrow a £10 note.
What year is it from?
Does it have the same features?

If you hold a note up to the light
you see a special watermark.

4 Look at the two serial numbers on one note.
  Are they the same or different?

5 Look at the serial numbers on two different notes.
  What do you notice?

6 Copy this table.

  Measure the height and width of a £5 note.
  Write them in your table.

  Look on a £5 note for the symbol to help
  people who are partially sighted.
  Copy the symbol into the table.
  Use the right colour.

  Do the same for a £10 note and a £20 note.

Note	Height	Width	Symbol
£5	cm mm	cm mm	
£10	cm mm	cm mm	
£20	cm mm	cm mm	

7 Design a new £10 note.
  Try to make a design that would be difficult to copy.

Coins and notes

a game you can play on your own or with one or two friends

You need: a dice
             a counter for each person
             about £100 in plastic £1 coins
             the game board made from worksheet **A3–7**
             about 15 token £10 notes (those you and your friends designed
             or some made from worksheet **A3–8**)

Follow the instructions on the game board.

Playing Scrabble®

The plastic squares in Scrabble are called **tiles.**

1 Ann puts down this word.
Work out her score by adding up
the points on the tiles.

2 Now it is Ben's turn.
He thinks about putting down these tiles.
What would his word score?

Remember to include this point in his score.

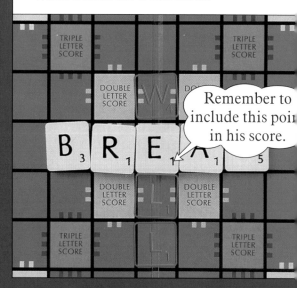

5 What would Ben score with CAKE?

A tile on this square gets its score multiplied by 3.

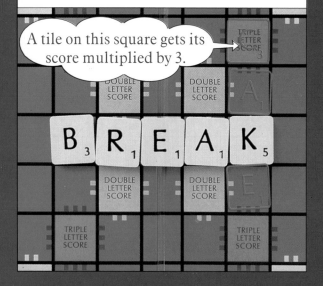

6 What is the score for this word?

3 Ben thinks he could get more points by putting down these tiles.
What would the word CLEAR score?

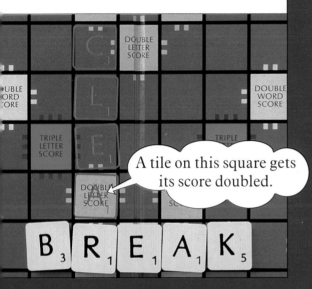

A tile on this square gets its score doubled.

4 Ben has a better idea.
What would his word score?

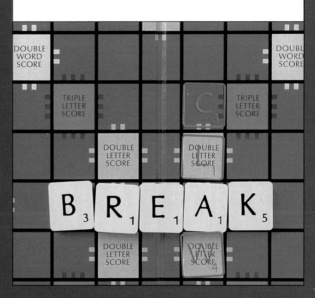

7 Later in the game Ben makes the word GATE.
What does it score?

Having a tile on this square doubles the score for the whole word.

8 Talk to your teacher about what Ben would score if he made the word GATES.

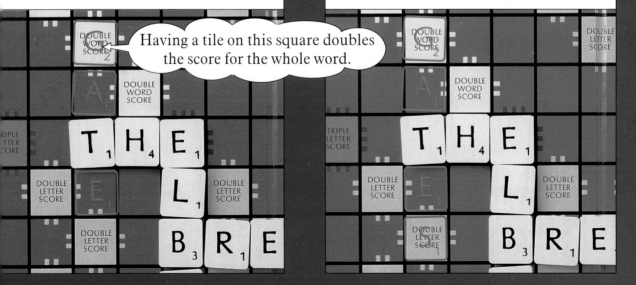

Scrabble® is the registered trademark of J. W. Spear & Sons PLC

◆ a.m. and p.m.

A When exactly do you mean?

I'll meet you at 8, Nicola.

OK, Kevin, at 8, then.

A1 Explain why Kevin and Nicola will not meet.

When we tell somebody a time they need to know what part of the day we mean.

We can say **morning**, **afternoon** or **evening**, or we can use **a.m.** or **p.m.**

We put **a.m.** after times in the **morning** and **p.m.** after times in the **afternoon** or **evening**.

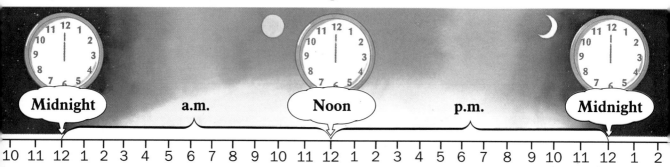

A2 Write each time using a.m. or p.m.

(a) Evening sport starts at 7:45.

(b) The Breakfast show is from 6:30 to 8:30.

(c) The late night film starts at 11:00.

(d) Afternoon theatre starts at 1:00.

(e)
(f)
(g)

A3 Write whether each of these times is
morning, **afternoon** or **evening**.

(a) 9:15 a.m.   (b) 1:30 p.m.   (c) 8 p.m.

(d) 3:30 p.m.   (e) 9:30 p.m.   (f) 6 a.m.

A4 Write these times in numbers. Use a.m. or p.m.

(a) Ten thirty in the morning   (b) Six fifteen in the evening

(c) One fifteen in the afternoon   (d) Nine thirty at night

B Using a clock

You can use a minute hand guide if you like.

It's a quarter to 2.
That's 1:45.
But it is the afternoon,
so I write 1:45 p.m.

B1 Write these times using a.m. and p.m.

(a) Half past two in the afternoon

(b)

(c) Quarter past nine in the morning

(d)    (e)

(f) Quarter to eleven in the evening

(g) Ten to ten in the morning

(h) Twenty-five to one in the afternoon

B2 12:30 a.m. is just after midnight.
When is 12:30 p.m.?

21

Calendar

Something to talk about

You can write the **first of October** in different ways.

1st October   October 1st   1 October   1st Oct

Think of some other ways. How do **you** write dates?

Days of the week

Dates in the month

October 1997

Sun	Mon	Tue	Wed	Thu	Fri	Sat
			1	2	3	4
5	6	7	8	9	10	11
12	13	14	15	16	17	18
19	20	21	22	23	24	25
26	27	28	29	30	31	

1   What day of the week is 1st October 1997?

2   What day of the week is 4th October?

3   What day of the week is 31st October?

4   Mandeep hires a cement mixer on Monday 6th October.
    It is due back the following Monday.
    What **date** is it due back?

5   Kim posts a card to her auntie on 22nd October.
    It gets there 3 days later.
    What **date** is that?

CAR BOOT SALE held here
first Sunday in the month

6   What **date** is the car boot sale in October 1997?

7   Doug gets his pocket money every Friday.
    How many lots of pocket money does he get in October 1997?

Angle art 2

patterns that make another good wall display

You need compasses, an angle measurer and felt-tip pens.

1 Put a dot near the centre of
a square piece of plain paper.
Set your compasses to 2 cm and
draw a circle with its centre on the dot.

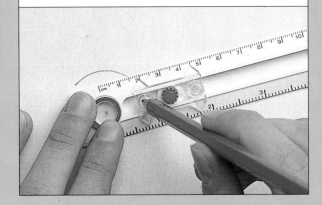

2 Set your compasses to 3 cm.
Draw another circle with
its centre on the dot.
Keep drawing circles that are
1 cm bigger each time until
you have filled the paper.

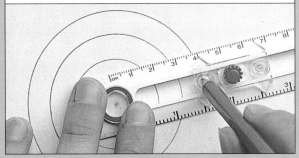

3 Put the centre of the angle measurer
on the centre point.
Mark off every 60 degrees all round.

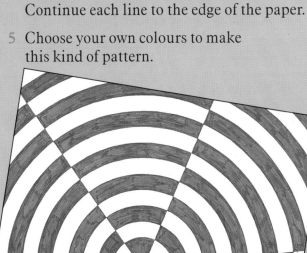

4 Draw a line from the centre to each of
the marks you have just made.
Continue each line to the edge of the paper.

5 Choose your own colours to make
this kind of pattern.

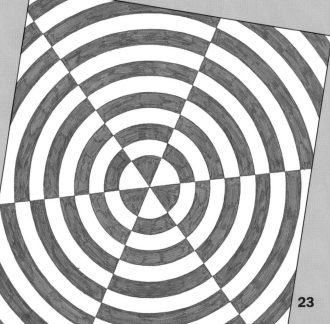

Cafe

A Helping in Mrs Owen's cafe

Kim and Jed get a holiday job serving in a cafe.

A1 Jed looks after the small tables.
2 customers can sit at each small table.

   (a) How many small tables are there?

   (b) How many customers could
      sit at the small tables **altogether?**

A2 Jed lays all the small tables.
Each small table needs these things.

    2 knives

    2 forks

    2 spoons

    5 sachets of tomato ketchup

    5 sachets of brown sauce

    5 sachets of salad cream

    10 sachets of sugar

   (a) How many knives will he need **altogether?**

   (b) How many forks will he need?

   (c) How many spoons?

   (d) How many sachets of tomato ketchup?

   (e) How many sachets of brown sauce?

   (f) How many sachets of salad cream?

   (g) How many sachets of sugar?

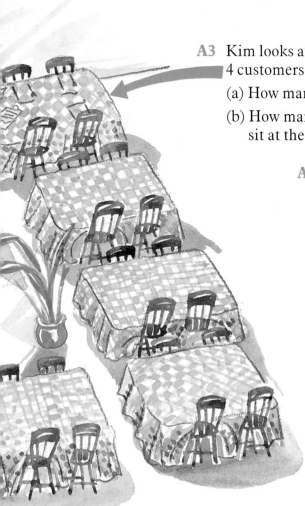

A3 Kim looks after the big tables.
4 customers can sit at each big table.

(a) How many big tables are there?

(b) How many customers could
sit at the big tables **altogether?**

A4 Kim lays these things on each big table.

> 4 knives
> 4 forks
> 4 spoons
> 9 sachets of tomato ketchup
> 9 sachets of brown sauce
> 7 sachets of salad cream
> 20 sachets of sugar

(a) How many knives will she need **altogether?**

(b) How many forks will she need?

(c) How many spoons?

(d) How many sachets of tomato ketchup?

(e) How many sachets of brown sauce?

(f) How many sachets of salad cream?

(g) How many sachets of sugar?

Mrs Owen does the cooking and the washing-up.
She stacks the plates in piles of 10.

A5 How many plates will there be in
(a) 5 piles? (b) 7 piles?

A6 She needs 20 plates at breakfast time.
How many piles is that?

A7 She has exactly 4 piles of plates before lunch.

(a) How many plates is that?

(b) She uses 34 plates. How many plates are left?

A8 She runs out of clean plates at tea time.
So she washes up 27 plates.

(a) How many piles of 10 can she make?

(b) How many plates will be left over?

B Calculate the cost

You may need coins.

Simon works out this bill.

Crisps	30p
Apple	35p

I do 30 + 35 on my calculator.
I get 65.
That means I pay 65p.

B1 Work out these bills on a calculator.
Give your answers in pence.

(a)
Crisps	30p
Squash	45p

(b)
Apple	35p
Bun	50p

(c)
Crisps	30p
Tea	60p

(d)
Apple	35p
Toast	55p

(e)
Toast	55p
Squash	45p

B2 This bill comes to more than one pound.
How much does it come to?

Cake	80p
Tea	60p

When we write amounts of money, people have agreed
to use either the £ sign, or the letter p, but not both.

£1·40 ✓   140p ✓   £1·40p ✗

B3 Write these amounts with a £ sign instead of a letter p.

(a) 150p   (b) 165p   (c) 142p   (d) 199p   (e) 250p

B4 Copy these and complete them.

(a) 200p = £_____   (b) 205p = £_____   (c) 95p = £_____

B5 Add up these bills.
Check with coins if you want to.

(a)
Crisps	30p
Coke	80p

(b)
Tea	60p
Apple	35p

(c)
Cake	80p
Coffee	65p

(d)
Cake	80p
Milkshake	95p

(e)
Toast	55p
Coffee	65p

c Don't mix pounds and pence!

What does my bill come to?

Salad £1·20
Coffee   65p

That's £66·20

No, that can't be right!

Be careful when you use a calculator to add up money.
You must add up in pence or add up in pounds.

You can do $\begin{array}{c} 120p \\ + 65p \end{array}$ or $\begin{array}{c} £1·20 \\ +£0·65 \end{array}$ but **not** $\begin{array}{c} £1·20 \\ + 65p \end{array}$

I see where I went wrong.
I pressed 1 · 2 0 + 6 5 =
and the calculator thought I meant £1·20 + £65.

C1 Add up these bills with a calculator.
Check with coins if you want to.

(a)
| Salad | £1·20 |
| Toast | 65p |

(b)
| Sandwich | £1·05 |
| Tea | 60p |

(c)
| Burger | £1·50 |
| Coffee | 65p |

Remember: when you are working in pounds on a calculator,
an answer like this 2.4 means £2·40.

C2 Add up these bills with a calculator.

(a)
Coffee	65p
Tea	60p
Omelette	£2·20

(b)
Apple	35p
Squash	45p
Crisps	30p

(c)
| Crisps | 30p |
| Eggs on toast | £1·15 |

(d)
| Milk | 80p |
| Salad | £1·20 |

D Cherry's cafe an activity to do with a friend

You need: the drinks cards (blue), the meals cards (green) and
the sweets cards (yellow) made from worksheets **A3–9**, **A3–10** and **A3–11**.

One person is the customer. The other is the waiter.
The customer chooses three cards, one of each colour.
The waiter writes out the bill on worksheet **A3–12** and adds it up.
The customer must check that the bill is right.

Take it in turns to be the customer and the waiter.

Make a kilogram

You need the weight cards made from worksheet **A3–13** .

Remember:

> 1 kilogram is 1000 grams.

A Working by yourself

Spread out the cards showing items of food and their weights.

A1 Pick out two items that add up to 1000 grams.
Write their names and weights.

A2 Write some other ways of getting a total of 1000 grams.

A3 Find the largest number of items that
add up to a total of 1 kilogram.

B Playing a game with a partner

Getting started

Shuffle the cards and deal three cards to each person.

Put the rest of the pack face down on the table.
Turn the top card over and lay it down next to the pack.

When it is your turn

Pick up a card from **either** pile.
Then choose one of your cards to put down on the face-up pile.
You should now have three cards again.

> You win if you have three cards that add up to a kilogram.

Vertical

A Judging vertical
a group activity

You need: map pins
a noticeboard
a plumb-line made from a piece of string
with a metal nut tied on one end

1 Put one pin near the top of the notice board.

2 At the bottom of the notice board each person tries to put a pin **exactly below** the top pin.

3 When everyone has had a go, hold the plumb-line against the top pin.

4 The winner is the person with the pin closest to the plumb-line.

Putting up wallpaper
Angie uses a plumb-line
to make sure that her first piece
of wallpaper is vertical.

Upright trees
Trees usually grow vertically,
even if they are on the side of a hill.

B How much does it lean?

You need an angle measurer.

Buildings are usually stronger if
their walls are vertical.

The famous tower at Pisa in Italy was
vertical when it was built.
But its foundations were not strong enough
and it began to lean.

1   Look carefully at your angle measurer.
    Each of these small spaces is one **degree**.

2   Set the red arrows to zero.

3   Put the angle measurer on the photo.

Check that the
zero line and
the hole fit
like this.

4   Turn the red arrows until they are
    on the red line on the photo.

B1   How many degrees is the tower leaning?

This pub in Staffordshire started to lean because
a salt mine under it made the ground weak.

1 Set the red arrows to zero again. Put your angle measurer on the photo.	2 This time turn the red arrows to the left and look at **this** scale.

Check that the zero line and the hole fit like this.

B2 How many degrees is the pub leaning?

B3 There are some drawings of leaning buildings on worksheet **A3–14**.
Use your angle measurer to find out how far each building is leaning.

The Hancock family

Fred Hancock

This family photo was taken on Fred's 70th birthday.

1 These are Fred's five granddaughters.

Hayley
born 1980

Helen
born 1979

Kate
born 1978

Claire
born 1982

Sarah
born 1984

(a) In what year was Hayley born?

(b) In what year was Kate born?

(c) Who is the older, Hayley or Kate?

(d) Who is the oldest of all five girls?

(e) Make a list of Fred's granddaughters, in order of their age, starting with the youngest.

2 These are Fred's grandsons.

(a) Who do you think is older,
    Mark or Simon?

(b) Look at when the granddaughters
    were born.
    Make a sensible guess which year
    Mark was born.

(c) What year do you think Simon
    might have been born?

3 Fred has three children – William, John and Mary.

(a) Who is older, William or John?

(b) Do you think Mary is older or
    younger than her brother John?

(c) Do you think Mary is older or
    younger than her brother William?

(d) What year do you think
    Mary might have been born?

4 Pete and Andrew are Fred's nephews.
   Who is older, Pete or Andrew?

5 Pete and Paula got married in 1984.

(a) What year did they have their first wedding anniversary?

(b) What year did they have their tenth wedding anniversary?

6 Fred remembers when Kate was born.
   The date was 3rd November 1978.

(a) What was the **date** of Kate's first birthday?

(b) What was the date of Kate's tenth birthday?

Challenge

Do you have an old photo of your family?
Try to find out what year everyone was born.

Have you ever wondered what year your teachers were born?
They might tell you if you ask politely.

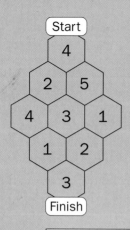

Honeycombs

This is a honeycomb.
Each cell has a number in it.
You draw a path by moving down through cells that touch one another.
You add the numbers as you go.

1 Steven has drawn a path.
What is the total of his numbers?

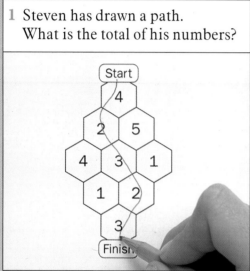

2 This is Narinder's path.
What is her total?

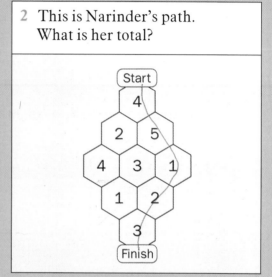

3 There are some more copies of this honeycomb on worksheet **A3–15**.
Draw some paths of your own.
Write the total each time.

Try to draw the path with the smallest possible total.
Label it 'smallest total'.

Draw the path with the largest possible total.
(Remember, you must move **down**, not from side to side.)
Label the path 'largest total'.

4 Worksheet **A3–16** has some copies of a different honeycomb.
Draw some paths.
Draw the path with the smallest possible total.
Draw the path with the largest possible total.

5 Worksheet **A3–17** has another honeycomb.
Draw some paths, including the one with the smallest total and
the one with the largest total.

We asked 100 teenagers

A Collecting the facts

You need a pie chart scale.

We asked 100 teenagers …

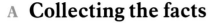

> Name something you would take swimming.

Our survey showed …

Swimming costume	50
Towel	28
Goggles	11
Other	11

> 50 out of the 100 teenagers said 'a swimming costume'.

> 11 out of the 100 teenagers gave other, less popular answers.

A number out of 100 is called a **percentage**.
So 50 **per cent** of teenagers said 'a swimming costume'.

A1 What percentage of teenagers said 'a towel'?

A2 What percentage of teenagers said 'goggles'?

A3 What percentage of teenagers gave other answers?

We can show these results on a **pie chart**.

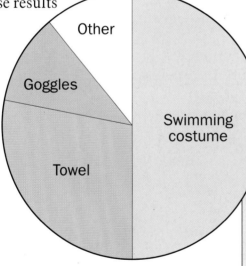

Use a **pie chart scale** to check that the percentages are correct.

T

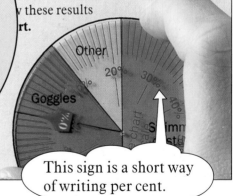

> This sign is a short way of writing per cent.

Next we asked the same 100 teenagers these.

1 Name something you listen to.
2 What is your best excuse for not having done your homework?
3 Name something you read.
4 Name something you brush.

A4 What answer would **you** give to each one?

B Measuring pie charts

Our survey gave us these results for question 1
on the previous page.

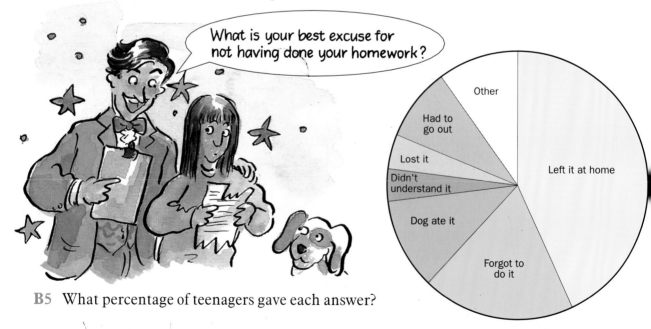

Name something you listen to.

Other

Radio

Walkman

Music

B1 (a) Which was the most popular answer?

(b) Use a pie chart scale to find out
how many teenagers gave this answer.

B2 How many teenagers said 'Walkman'?

B3 What percentage of teenagers said 'Radio'?

B4 What percentage of teenagers gave other answers?

Our survey gave us these results for question 2.

What is your best excuse for
not having done your homework?

Other

Had to
go out

Lost it

Didn't
understand it

Dog ate it

Left it at home

Forgot to
do it

B5 What percentage of teenagers gave each answer?

c Drawing your own pie charts

The survey gave us these results for question 3.

Name something you read.

Book	69
Magazine	19
Newspaper	7
Other	5

C1 Draw a pie chart of these results.

Here are the results for question 4.

Name something you brush.

Hair	75
Teeth	16
Dog	3
Horse	3
Other	3

C2 Draw a pie chart to show these results.

Challenge

In your group interview 100 teenagers and show the results you get. You can use some questions from this chapter or some of your own.

First talk to your teacher about how to share out the work and how to collect all the results together.

Four cubes

You need about 50 multilink cubes, and 13 pieces of sticky label.

Each of these models is made from four cubes.
Build all the models with the cubes.
Label each one with its letter.

1 Some of the models are really the same shape, but
turned a different way round.
For example, C and K are the same shape.
Find some other models that are really the same shape.

2 One model is not the same shape as any of the others.
Which one is it?

3 Try to make a different model by putting four cubes together.
Show your model to your teacher.

24-hour times

a.m. and p.m. system	24-hour system
12 midnight	00:00
1:00 a.m.	01:00
2:00 a.m.	02:00
3:00 a.m.	03:00
4:00 a.m.	04:00
5:00 a.m.	05:00
6:00 a.m.	06:00
7:00 a.m.	07:00
8:00 a.m.	08:00
9:00 a.m.	09:00
10:00 a.m.	10:00
11:00 a.m.	11:00
12:00 noon	12:00
1:00 p.m.	13:00
2:00 p.m.	14:00
3:00 p.m.	15:00
4:00 p.m.	16:00
5:00 p.m.	17:00
6:00 p.m.	18:00
7:00 p.m.	19:00
8:00 p.m.	20:00
9:00 p.m.	21:00
10:00 p.m.	22:00
11:00 p.m.	23:00
12 midnight	00:00

Morning · Afternoon · Evening

We use a.m. and p.m. so we do not mix up times in the morning with times in the afternoon or evening.

Another way of avoiding confusion is to use the **24-hour** system of telling the time. Timetables for buses and trains use this system. Some digital clocks and watches use it too.

1 Use the diagram to change these times into the 24-hour system.

(a) 3:00 p.m. (b) 7:00 a.m. (c) 10:00 p.m.

(d) 5:00 p.m. (e) 4:00 a.m. (f) 6:30 p.m.

Times in the 24-hour system always have four digits. So 7 o'clock in the morning is **07:00**.
Also, we **don't** write a.m. or p.m. in the 24-hour system. Look at your answers to check you got these things right.

2 Change these into 24-hour times.

(a) Four o'clock in the afternoon

(b) Ten o'clock in the morning

(c) Eight o'clock in the evening

(d) Half past one in the afternoon

(e) Twelve noon

(f)

3 Write these 24-hour times as a.m. or p.m. times.

(a) 23:00   (b) 08:00   (c) 21:00

The 24-hour system is often called the **24-hour clock**. But you don't need a special clock to use it.

39

Ted Sharp works at the information office at the railway station.
He uses this rule to turn p.m. times into 24-hour times.

Add 12 hours to the time.

The time is nearly 2:00 p.m.
He adds 12 hours and gets 14:00.

You've just got time to catch the 14:00 to York, madam.

4   Use Ted's rule to turn these times into 24-hour times.

(a) 6:00 p.m.

(b) One o'clock in the afternoon

(c) 1:30 p.m.

(d) Seven o'clock in the evening

(e) 10:30 p.m.

(f) Half past eight in the evening

5   Write these times as a.m. or p.m. times.

(a) 08:00   (b) 16:30   (c) 20:30

Kitchen scales

A Pounds and kilograms

I've got a pound of apples.

I've got a kilogram of apples.

Sometimes we weigh things in pounds and ounces.
Other times we use a different system, and
weigh things in kilograms and grams.

Many shops that used to use pounds and
ounces now use kilograms instead.

When you were born they might have
weighed you in pounds or kilograms.

Apples
52p
a kilo

Apples
25p
a pound

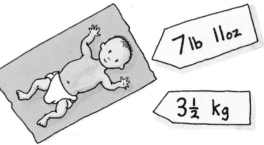

7 lb 11oz

3½ kg

A1 Do you know what you weighed when you were born?
Were you weighed in pounds or kilograms?

A2 Look at these two weights.
Which of them do you think is heavier?

lb is a short way
of writing **pound.**

kg is a short way
of writing **kilogram.**

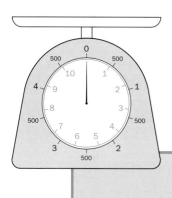

B The numbers on the scales

Kitchen scales often have two lots of numbers on them.
These scales will weigh in kilograms and grams (the **black** numbers) or in pounds and ounces (the red numbers).

In the rest of this chapter we will use kilograms and grams.

1 kg = 1000 g

A kilogram is made up of a thousand grams.

This big pack of sultanas weighs a kilogram.

This is a gram of sultanas.

This bag of rice weighs 2kg.

I put it on the scales and this is what they look like.

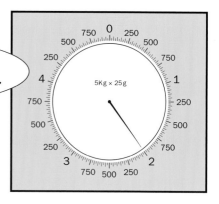

B1 I have put a pack of spaghetti on. What does it weigh?

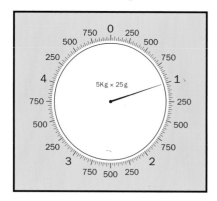

B2 I have put some potatoes on. What do they weigh?

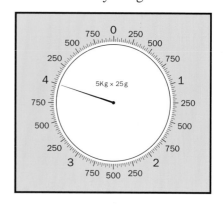

B3 How many grams are there in half a kilogram?

B4 Sally has put a tub of margarine on here.
What does it weigh?

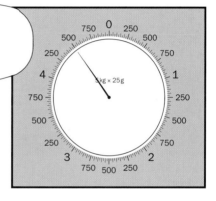

I have put a parcel on these scales.
It weighs 4 kg 500 g.
Another way of saying this is 4½ kg.

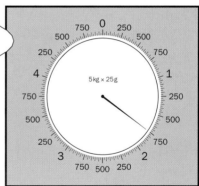

Now do question **B5** on worksheet **A3–18** .

My rabbit's food weighs 1 kg 750 g.
This is what the scales look like.

Now do question **B6** on worksheet **A3–19** .

c Different scales

Not all weighing scales are the same.
Think of different kinds of scales
that you have seen.
Try to draw some of them.

Review: Book A2 and pages 4 to 17

A Multiplication

A1 Write the answers to these.

(a) 4×3 (b) 5×3 (c) 2×5 (d) 2×4

B Rulers

This bicycle spanner fits nuts of different sizes.

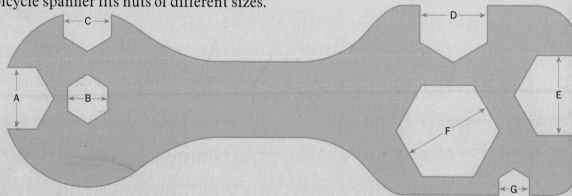

B1 Measure the openings on the spanner in centimetres and millimetres.
Write your answer like this.

A = 1cm 6mm
B =

B2 Here are four bicycle nuts.
Measure their sizes in **millimetres**.

(a)   (b)   (c)   (d)

B3 Write which opening on the spanner fits on to each nut.

C Money

C1 Jo has three £10 notes and six pound coins.
How much money does she have altogether?

C2 Kerry has four £10 notes.
She swaps them all for £1 coins.
How many coins does she get altogether?

C3 Lee has four £10 notes and four pound coins.
How much money does he have altogether?

Review: pages 11 to 22

A Halving recipes

These are the quantities for a blackberry pie big enough for 8 people.

For the **filling**: 400g blackberries
120g sugar

For the **pastry**: 300g flour
200g butter
100g sugar

Cathy doesn't have enough blackberries for the full recipe, so she decides to halve all the quantities.

A1 Write the quantities that Cathy will use.

A2 How many people will Cathy's pie be big enough for?

B a.m. and p.m.

B1 Write something that you might be doing at each of these times on a school day.

(a) 7:45 a.m.   (b) 1:30 p.m.   (c) 4:20 p.m.

(d) 2:00 a.m.   (e) 10:15 a.m.   (f) 8:30 p.m.

B2 Write these times using a.m. or p.m.

(a) Half past three in the morning

(b)

(c) Quarter past seven in the evening

C Calendar

February 1997						
Sun	Mon	Tue	Wed	Thu	Fri	Sat
						1
2	3	4	5	6	7	8
9	10	11	12	13	14	15
16	17	18	19	20	21	22
23	24	25	26	27	28	

C1 What day of the week is 21st February 1997?

C2 What date is the first Friday in February 1997?

C3 How many Fridays are there in that month?

C4 How many Saturdays are there?

C5 Nathan wants to go on a coach trip on Saturday 15th February. He has to pay on the Saturday before that. What **date** must he pay on?

A Angle art

You need compasses, an angle measurer and felt-tip pens.

1 Put a dot near the centre of
a piece of plain paper.
Set your compasses to 2cm and
draw a circle with its centre on the dot.

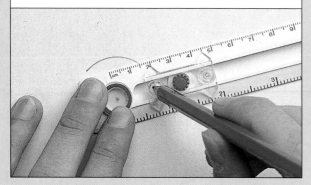

2 Draw more circles that are
1cm bigger each time until
you have filled the paper.

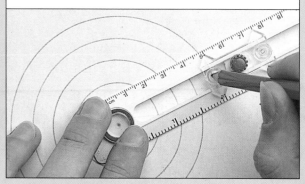

3 Put the centre of the angle measurer
on the centre point.
Mark off every 10 degrees all round.

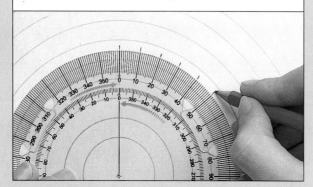

4 Draw a line from the centre to
each of your marks and continue to
the edge of the paper.
Make a pattern like this.

B Cafe

SPECIAL PRICE MEAL
Burger, chips, beans and soft drink
ONLY £2·95

B1 These are the normal prices.

(a) Add up the normal cost of the meal.

(b) How much do you save when
you buy the special price meal?

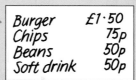

Burger	£1·50
Chips	75p
Beans	50p
Soft drink	50p

Review: pages 24 to 37

A Cafe

A1 Five children sit at each table for a party.

(a) How many tables are there?

(b) How many children will there be if all the chairs are full?

A2 Each table needs these things.

    1 salt pot
    1 pepper pot
    5 knives
    5 forks
    5 spoons
    10 sachets of tomato sauce

(a) How many salt pots will be needed altogether?

(b) How many pepper pots will be needed?

(c) How many knives will be needed?

(d) How many forks will be needed?

(e) How many spoons will be needed?

(f) How many sachets of tomato sauce will be needed?

B Make a kilogram

These packets have their weights written on.

B1 Choose 3 packets that weigh 1 kilogram altogether. Write their colours.

B2 Choose 4 packets that weigh 1 kilogram altogether. Write their colours.

C Percentages

You need a pie chart scale.

This pie chart shows how 100 people spent their May bank holiday.

C1 Use a pie chart scale to answer these questions.

(a) How many people stayed at home?

(b) How many people went to visit friends?

(c) How many people went to the seaside?

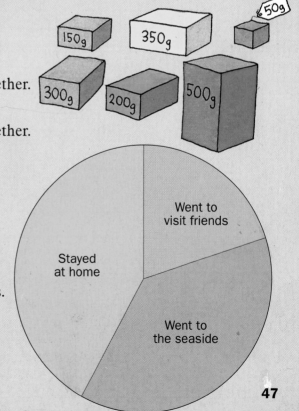

47

Review: pages 32 to 43

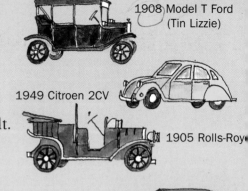

1908 Model T Ford
(Tin Lizzie)

1949 Citroen 2CV

1905 Rolls-Roy•

A Dates and ages

Here are some old cars with the years they were built.

A1 Which is the oldest car?

A2 Which is the newest car?

A3 Write the names of the cars in order of their age, starting with the oldest.

A4 Which car was built 20 years after the Citroen 2CV?

A5 How many years after the model T-Ford was the Morris Minor built?

1969 Ford Capri

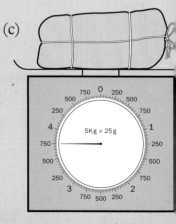

1948 Morris Minor

B Percentages

You need a pie chart scale.

B1 This is how a hundred people said they spent one Saturday afternoon.

Draw a pie chart to show this information.

Went to football match	35
Went out shopping	30
Stayed at home	22
Other	13

C 24-hour times

C1 Write these 24-hour times as a.m. or p.m. times.

(a) 10:00   (b) 14:30   (c) 21:00   (d) 05:30

D Kitchen scales

D1 How many grams are there in a kilogram?

D2 How much do these parcels weigh?

(a)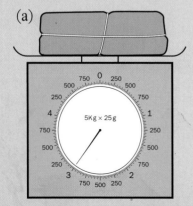

(b)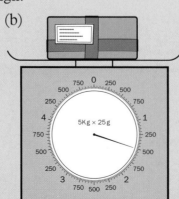

(c)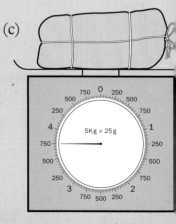